ISBN 978-1-332-85994-8
PIBN 10253869

This book is a reproduction of an important historical work. Forgotten Books uses
state-of-the-art technology to digitally reconstruct the work, preserving the original format
whilst repairing imperfections present in the aged copy. In rare cases, an imperfection in
the original, such as a blemish or missing page, may be replicated in our edition. We do,
however, repair the vast majority of imperfections successfully; any imperfections that
remain are intentionally left to preserve the state of such historical works.

English
Français
Deutsche
Italiano
Español
Português

www.forgottenbooks.com

Mythology Photography **Fiction**
Fishing Christianity **Art** Cooking
Essays Buddhism Freemasonry
Medicine **Biology** Music **Ancient
Egypt** Evolution Carpentry Physics
Dance Geology **Mathematics** Fitness
Shakespeare **Folklore** Yoga Marketing
Confidence Immortality Biographies
Poetry **Psychology** Witchcraft
Electronics Chemistry History **Law**
Accounting **Philosophy** Anthropology
Alchemy Drama Quantum Mechanics
Atheism Sexual Health **Ancient History**
Entrepreneurship Languages Sport
Paleontology Needlework Islam
Metaphysics Investment Archaeology
Parenting Statistics Criminology
Motivational

Rideau Waterway Guide

BY BOAT AND CAR
THROUGH
THE RIDEAU LAKES AND THE RIDEAU CANAL

COMPLETE WITH MAPS
AND TOUR INFORMATION

One Dollar and Twenty Five Cents

ROBERT HAIG PUBLISHING COMPANY
OTTAWA, ONTARIO

CONTENTS

HOTEL KENNEY, JONES FALLS, ONTARIO

ROBERT HAIG PUBLISHING COMPANY

Suite 3   186 Bank St.

Ottawa, Ontario.

2

The picturesque Rideau Canal is one of Canada's truly great vacation attractions. Forming an inland water route for 123 miles from historical Kingston to Ottawa, our Nation's Capital, this scenic cruiseway is plied by thousands of large and small craft each happy summer.

As well as being enjoyed by the many fun seekers who sail and splash in its waters the Rideau Canal is easily accessible by car and the many locking stations provide wonderful opportunities for landlubbers to come and laze away a quiet afternoon or evening in the incomparable beauty of this part of Eastern Ontario.

Not only does the canal provide an enviable vacation pursuit it serves also to link together many of the finest fishing lakes to be found in all Ontario. Mention names like Opinicon and Whitefish, or Loughborough and Sand and Newboro and Big Rideau to those who have visited here and you will be regaled with tales of happy holidays spent in this carefree country.

The many typical Eastern Ontario towns and hamlets like Perth, Smiths Falls, Battersea, Seeleys Bay, Westport, and Merrickville, to name but a few, are the friendliest communities imaginable and will make your visits to these vacation spots as informal and pleasurable as you yourself want them to be.

Be one of the many to enjoy the fascinating, and pleasure filled, Rideau Waterway.

Eastern Ontario

ACCESS

Flanked by the State of New York to the south and the Province of Quebec to the east and north, Eastern Ontario is situated in a most favourable position to the eighty two million people who reside within a 500 mile radius, or one day's drive, from this vacationland.

From the United States access by car can be made by any of the excellent roads leading to the St. Lawrence River Valley between Alexandria Bay and Rooseveltown. Three high rise bridges cross the St. Lawrence River. The one at Alexandria Bay affords a wonderful view of the Thousand Islands as it leads to Hwy 401 near Ivy Lea. A second bridge stretches from Ogdensburg to Prescott and a third winds over the great Seaway project between Massena and Cornwall.

Within Ontario, Hwy 401 traverses the southern part of the province from Windsor, near Detroit, to east of Cornwall at the Quebec border. Other toll free highways that lead to Eastern Ontario are Hwys 2, 7 and 17 from the West and Hwys 2 and 17 from the East. Within Eastern Ontario almost a dozen highways criss-cross between all the major areas. Of particular importance to viewing the Rideau Waterway are Hwys 15, 43 and 16. Locate these on the map on the rear cover of this guidebook for easy recognition.

Entrance by cruiser boat to the Rideau Canal system is made at the Cataraqui River at Kingston or one may sail up the Ottawa river to Ottawa, our Nation's Capital, to enter via the Giant's Staircase, a series of eight locks lifting up to the very centre of the city.

TOPOGRAPHY

Rolling farmlands predominate much of Eastern Ontario. The rich soil produces crops of excellent yields and highest quality. Numerous small cheese factories, many of which enjoy a world wide reputation for their prize winning cheeses, are located throughout the valley.

From north of Gananoque to as far as Perth and then north-west through the Upper Ottawa Valley and the Madawaska Valley the surface features change quickly from gentle hummocks and vales to

Parliament Buildings, Ottawa

5

the more .rugged, sharply defined hills of the Precambrian Shield. Classed as the oldest rock structure in the world, this great mass of stone covered only thinly with earth, was formed perhaps a billion years ago. Known in New York State as the Adirondack Mountains and hereabouts as the Gatineau Hills and the Laurentian Mountains it stretches back over half of Canada to as far as Labrador in the east and the Mackenzie River in the north. To-day one can trace the leading edges of this massive rock shield at Battersea, Jones Falls, Big Rideau Lake and along many sections of Hwy 15. The many beautiful lakes that lie between Perth, Kingston and Brockville are a happy consequence of this act of Nature. The Thousand Islands in the St. Lawrence River are the very tips of great peaks depressed by the forces of mile thick ice fields that lay over the area count-less centuries ago.

POPULATION

Eastern Ontario has a population of about 830,000 representing 4.30% of the total Canadian population. It enjoys a growth rate of 28% over a 10 year period. Almost one half of the people live in Metropolitan Ottawa. The next largest community is Kingston pop. 55,000 followed closely by Cornwall pop. 43,000. The larger towns are Hawkesbury (8800), Prescott (5200) Brockville (19,000) Gananoque (5200) Perth (5700) Smiths Falls (10,000) Carleton Place (4800) Arnprior (5700) Renfrew (8500) and Pembroke (17,000).

Smiths Falls is the geographical centre of the Rideau Canal system which is 123½ miles long from Kingston to Ottawa. Some mileages to Smiths Falls are Cornwall 76; Syracuse, N.Y. 140; Montreal 200; Rochester N.Y. 215; Albany N.Y. 215; Toronto 215; Binghampton N.Y. 230; Scranton, Penn, 290; Harrisburg, Penn. 395; New York City 415; Cleveland, Ohio 465.

HISTORY

Bounded on the north and the south by two great rivers, the Ottawa and the St. Lawrence, that enabled early voyageurs like Champlain, Dollard des Ormeaux, De la Salle, Radisson, Franklin, Frobisher and many others to paddle and portage their way to the vast inlands of North America, much of the development of Eastern Ontario was for long nothing but a slow increase in the number of scattered homesteads in the wilderness. Although Montreal was founded in 1643, and Kingston in 1673, almost two centuries passed before a respectably sized settlement could be found along the St. Lawrence between Kingston and Montreal.

The unfortunate War of 1812-14 between Britain and the United States spurred development along the St. Lawrence but only to keep the supply lines open between Toronto and Montreal. Continued threats of invasion from the U.S. necessitated a route safer than the the St. Lawrence could provide and thus the great Rideau Canal system of locks, canals and natural water routes, that we refer to as the Rideau Waterway, was developed so that safe military passage could be made from Kingston to Montreal via the canal and the Ottawa River. Opened in 1832, the Rideau Canal was responsible for the founding of Bytown, now Ottawa, and for the accelerated growth of communities like Smiths Falls and Perth.

The selection by Queen Victoria in 1857 that Ottawa be the Capital of the newly united provinces of Upper and Lower Canada did much to spur the expansion of this beautiful city. For the remainder of Eastern Ontario, however, it was not until much later, in 1954, that a major development materialized to profoundly affect the neighboring communities. This was the advent of the St. Lawrence Seaway, a joint project between the United States and Canada that took five years to complete. Some 10,000 persons were displaced to make way for the flooding of roads and of abandoned cellar sites where their homes had once stood. The tourist attraction qualities of the dams, locks and new parks have helped make this area prosper. Although the industrial potential of Seaway Valley is tremendous in scope it may be some years before this region achieves the growth rate it is so capable now of handling.

TOURISM

It is in the field of tourism that Eastern Ontario has shown greatest progress. Year after year an estimated two million persons travel by car, boat, train and plane to frolic and sightsee in this great vacationland of contrasts. Having an enviable reputation for fishing and hunting, with almost countless lakes and bays for boating and swimming, the area is extremely popular with outdoors- men of both types, those who return to snug accommodation at night and those who choose to camp out under the stars.

The region offers almost every form of sightseeing interest and holiday fun. From boating, hiking, fishing, touring and swimming to

Sunset Ceremony, Old Fort Henry, Kingston

Eastern Ontario...

picnicing under massive shade trees, there is something for everyone. Some major attractions for all to see are the parading of guardsmen in their colourful uniforms at Ottawa, and at Old Fort Henry in Kingston, The 1000 Islands, Upper Canada Village, the Seaway Project, the Bonnechere Caves and the many beautiful parks and playgrounds that abound everywhere. The Rideau Waterway encompasses many of these top-rate vacation spots.

ACCOMMODATION

Plentiful accommodation to suit all tastes and all pocketbooks is available throughout Eastern Ontario. From deluxe hotels and motor hotels to motels, cabins and cottages and the homey lodge a hospitable welcome greets travellers from far and near. There is generally adequate trailer and camping facilities provided for outdoors living, some operated by private enterprise, others by the Ontario Government. An excellent standard of sleeping and eating facilities is maintained by the operators of the many establishments. Marinas and docking facilities are located along the Rideau Canal but boaters are cautioned to plan ahead well and wisely to ensure they are not caught between sets of locks where tie up and accommodation conveniences are not provided. Page 71 of this guide lists all marina and docking facilities throughout the canal system.

More than 1050 accommodation establishments are licensed by the Department of Tourism and Information to operate in Eastern Ontario. These provide nightly sleeping accommodation for almost 40,000 vacationers. About 360 hotels and motels operate on a year round basis. The remaining 690 establishments are primarily lodge and cottage resorts open generally from early May to late October. Approximately 100 registered camping sites offering a variety of facilities are located in the main areas.

For information on any aspect of vacationing in Eastern Ontario write to:

The Department of Tourism and Information,
Queen's Park,
Toronto 2, Ont.

or to the Chamber of Commerce of the municipality in which you are interested.

The **Flight** of 8 **Locks, Ottawa**

10

The Rideau Canal

The Rideau Canal was built during the six year period 1826-32 by the British Government to provide an alternate and safer route for the transport of troops and military supplies from Montreal to Kingston.

At the turn of the 18th century and well into the 19th century the port of Kingston was an important military centre for the defence of Upper Canada, as Ontario was then called. It was imperative therefore to maintain a safe supply route from Montreal since the isolation of Kingston by American forces would assuredly have meant the loss of York, now Toronto, and many other British communities along Lake Ontario. As early as 1783 the British Government had despatched a Lt. French to explore the inland water route — from where Ottawa now stands, then a wilderness, down to the St. Lawrence River. His report is the first written record of essentially the same path the Rideau Canal was to follow years later.

The War of 1812-14 between Britain and the United States placed increased emphasis upon the military dependence of the St. Lawrence River supply line and it was during this period that Sir George Prevost, commander of the British forces, initiated a request complete with a proposed plan that consideration be given to building a canal to by-pass this vulnerable route. In 1816 a further excursion into the still relatively unknown Rideau route was made by a Lt. Jebb who completed the entire trip by water from Kingston to where the Rideau River discharges into the Ottawa River. However, it was not until 10 years later in 1826 that the British Government took a decision to construct the Rideau Canal.

The military engineer assigned the responsibility of completing this project was Lt. Col. John By of the Royal Engineers. A soldier with a distinguished record, and having previously served for nine years in Canada at Quebec, from 1802-11, he landed again at Quebec on May 30, 1826 and in late summer journeyed to Montreal to commence hiring contractors and arranging for supplies. His first visit to the junction of the Rideau and Ottawa rivers, across from which stood the little settlement of Hull, was made on September 21, 1826. Within five days the location of the entrance for the first set of locks was decided upon and the building of the Rideau Canal was laboriously begun.

The story of the building of the Rideau Canal is a fascinating one. Litèrally carved out of swamps, forests and solid rock, in strange and virtually uninhabited country where fever ran rampant in summer and freezing cold prevailed in winter, the spirits of the military engineers and the contractors remained undaunted throughout the entire six year period. About the only hazard they did not have to face was an attack from the Americans. To this day the canal has seen only one military uprising, that of the Rebellion of Upper Canada in 1837-38.

Starting at Ottawa, the preliminary construction was to raise a bridge at Chaudiere Falls to provide access from Hull to the site of the entrance locks. Land was cleared here and two small villages, Upper and Lower Town were laid out for use of the workers. Thus did the beautiful city of Ottawa have its honourable beginning. Col. By built a humble house near where the Chateau Laurier Hotel now stands, a workshop and headquarters building was erected by the side of the lock site and two wharves were completed at Entrance Bay.

Meanwhile a surveying team under John MacTaggart penetrated inland for five miles to Hogs Back. The bush was so thick and the swamp at where Dows Lake now stands so difficult to wade through it made surveying impossible and they had to wait until winter came to freeze the ground sufficiently to make it reasonably passable.

The first winter was occupied in making plans for the first set of locks, in hiring contractors, arranging for supplies and administering the many details of this sizable party. In May of 1827 Col. By journeyed the entire Rideau route to Kingston, suffering many inconveniences and dangers. Being a practical and well qualified engineer he made many decisions on this first survey which remained unchanged for the next five years.

Work on the first set of locks progressed satisfactorily. All labour was done manually as there were no steam shovels or bulldozers in those days. One dug with a pick and excavated with a shovel, hauling away the rubble in a wooden wheelbarrow. Oxen were used to pull heavy loads such as the handsome limestone blocks quarried from this very site and only simple block and tackle were employed to raise or lower these blocks into place. Rock was removed by the tedious process of hand boring holes into which gunpowder was poured to blast the rock, and too often the workers, out of existence. Lock gates were made by blacksmiths using English flat iron and the necessary iron castings were brought up from foundries in Montreal. Local timber was in plentiful supply and was worked by skilled carpenters. Particular mention should be made of the excellent work done by the stonemasons throughout the entire canal system. Every lock and dam has been expertly laid with precisely measured hand hewn granite or limestone blocks, each carefully grouted, that assuredly will withstand the elements for centuries to come.

The work was dangerous and exhaustive and many workers died from swamp fever, especially in the Jones Falls district, before the canal was completed. Col. By himself fell victim to the fever but fortunately rallied through to continue his task. It is some measure of indication of the administrative qualities of the military officers, and of the senior contractors too, that no threat of strike or refusal to work was ever uttered by the labouring forces.

During the year 1827 the entire area from Ottawa to Kingston was surveyed and the selection of the final route for the canal was decided upon. In this and the following year contractors were engaged to commence work on many of the locks and dams at the twenty three stations along the canal. Upon Col. By and his assistants fell also the task of settling the purchase of needed properties by the many falls where small mills had been established, and along other parts of the route where dams or locks were to be built.

Progress in 1828 was marred by a violent attack of a fever that prevailed throughout most of Upper Canada. Work was stopped at many stations until the pestilence subsided. Throughout 1829 and 1830 the speed of the many individual projects increased as experience was gained and more labour was brought in. The set of eight locks at Ottawa were completed amidst much celebration and in 1831 another extensive project, the arched dam and the four massive locks at Jones Falls were finished in their entirety as were the majority of other lock stations throughout the canal.

By the spring of 1832 the entire Rideau Canal project was completed and in working order. On May 24, 1832 Col. By and a party of invited officials boarded a boat at Kingston and sailed the 123.5 miles to Bytown to formally open the Rideau Canal.

Today the incomparable Rideau Canal is much the same as the day it was opened. If we disregard for a moment the presence of new railway and road bridges and close our eyes to the cottages and houses lining the shores, we can literally step back the 133 years separating us from Col. By and his officers and associates and better appreciate the tremendous achievements they performed.

By Car to all 49 Locks

Even if you do not posess a boat, come by car to enjoy the thrill of seeing pleasure craft locking through.

During the summer months the locks are in operation all day to about 8.30 p.m. Be sure to take your camera, a fishing rod and a picnic basket.

Nicholson's Locks, near Merrickville

OTTAWA - LOCKS 1-2-3-4-5-6-7-8

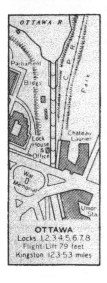

OTTAWA
Locks 1,2,3,4,5,6,7,8
Flight-Lift 79 feet
Kingston 123-53 miles

Between The Parliament Buildings and The Chateau Laurier Hotel.

If you are a stranger to Ottawa we suggest you ascertain in advance the best means of gaining access to these locks by car. Many visitors consider it best to leave their automobile at their motel or hotel and take a bus or taxi to Confederation Square from which it is but a few moments pleasant walk down to the locks.

In an incomparable setting of beauty and dignity, this majestic flight of eight locks rises in simple harmony with the rocky walls from which it was fashioned for a lift of 79 feet up to the canal channel above.

Best seen from the level of the Ottawa River, preferably from the sightseeing boat that plies here regularly, the magnificence of this single structure is a lasting tribute to the fortitude and skill of the Superintending Engineer, Lt. Col. John By, Royal Engineers. All who travel the Rideau Canal and know the fascinating history of the building of this great inland water route will want to pause here and reflect upon the everlasting achievement of this great man and of his intrepid team of soldiers and contractors.

Half way up the side of the locks may be seen a large stone building with black trim windows. Built in 1827 this building served as a headquarters for Col. By from which he directed and supervised the construction of the Rideau Canal. It has now been leased by the Historical Society of Ottawa and much space inside is devoted to the display of valued objects dating back to the early days of the canal. The gracious ladies attending this museum during the summer months give freely of their time to present to visitors interesting accounts of their collections. May we suggest you invest a few moments, and a few pennies admission, for a tour you will long remember. The museum is open Monday to Saturday 2 to 5 PM, Wednesdays 12 to 5 PM, from mid May to end September.

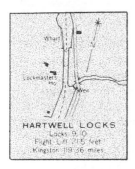

HARTWELL LOCKS
Locks 9, 0
Flight Lift 21.5 feet
Kingston 119.36 miles

Adjacent to Col. By Drive, between Hogs Back and Dows Lake, in Ottawa.

These two locks blend so quietly into the embankment alongside the road one can be excused for driving by without being aware of their existence. It is almost impossible nowadays to realize this beautiful setting was, at the time the canal was constructed, the edge of a dank, murky swamp stretching from Dows Lake, or Dows Great Swamp as it was then called, down past Bronson Avenue to the Rideau River. And it is difficult too for us to appreciate it took the early surveyors five days to hack their way from the start of the canal through a dense forest and somehow to wade this swamp to journey to Hogs Back, a route they had to travel many times over.

A fine view across to Carleton University is provided from the elevated vantage point of the upper of the two locks. The channel winds gracefully through this very attractive section of Ottawa for one half mile to Dows Lake thence the last few miles to the Ottawa River, imparting a most continental flavour and adding much to Ottawa's charm.

We had at first classified this set of locks as one of the few in the system without a dam but the Superintending Engineer for the Rideau Canal, Mr. L.W. Clark, kindly mentioned that near the upper lock is a weir. This weir serves for draining the canal channel up to Hogs Back to permit maintenance and repairs, a continuous requirement due to poor soil conditions affecting the foundations. Behind the weir is a tunnel through which the water flows underground alongside the locks to re-enter the canal near the gates of the lower lock.

HOGS BACK - LOCKS 11-12

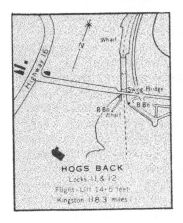

¼ mile east off Hwy 16 at Hogs Back, within Ottawa.

At this point the Rideau Canal and the Rideau River diverge to continue parallel routes for the next five miles, dropping 115 feet to the Ottawa River and the end of their courses.

The dam here had to be constructed in almost impenetrable wilderness and swamp and many engineering and building problems had to be overcome, not always without disaster and failure, before the turbulent waters surging down the rocky gorge could be brought under control. Taking almost three years to complete there is today but little apparent evidence of the tribulations which beset the engineers and contractors as they attempted to broach the raging waters with rock and earth quarried from the side of what is now Mooneys Bay. The water which flows over the weir under the stone bridge seems powerful enough as it rushes down this rocky chasm but it must surely be of only minor significance when compared to the seething cauldron prevalent before the building of the dam. Such is the quiet scene today that many people driving on the stretch of road between the swing bridge and the refreshment booth do not realize they are crossing a dam; nor are they aware the rubble that can be seen at the bottom of the gully on the north side and continuing down the river bed is all that remains of the third disastrous attempt to construct this dam.

The two locks nestle inconspicuously on the western bank in a small area of well cared for lawns and gardens. The channel continues downstream for one mile to Hartwells, the next set of locks.

The Hogs Back region is a popular recreation centre for boating, fishing, lounging in beautiful Vincent Massey Park overlooking the Rideau River, sunning on the sandy beaches at Mooneys Bay or sitting by at night watching the moonlit waters.

BLACK RAPIDS - LOCK 13

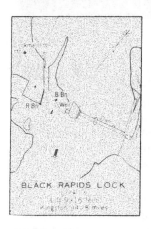

BLACK RAPIDS LOCK

1/3 mile east off Hwy 16, 2¼ miles south of Ottawa City limits.

Readily seen from the highway this attractive lock is another favourite haunt for Ottawans seeking an evening or a week-end in the quiet countryside. Only 10 miles from the centre of our Nation's Capital it often appears as remote as if it were in a distinctly subdued resort area. As is usual up and down the canal the indefatigable fisherman is to be found trolling or casting for a pickerel or bass. The well tended grounds form a perfect park for viewing this lovely waterway scene where the long dam with its white curtain of water curves gracefully away from the lock to the opposite bank of the river.

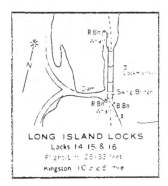

LONG ISLAND LOCKS
Locks 14 15 & 16
Flight:Lift 25·33 feet
Kingston 10 ½ ≤ 2 mile

½ mile west off River Road 2.4 miles north of Manotick. River Rd. runs parallel to Hwy 16 and may be reached by crossing the river at Manotick or driving out from Ottawa via Riverside Drive, keeping to the right of Uplands Airport. The sideroad is identified by a green pointer sign marked LONG ISLAND RD.

A not too well known site it is however especially popular with those who come here frequently to cool in the shade or frolic on the water. The three locks step down gracefully in a narrow channel by the river providing a change in water level of over 25 feet. The channel continues downstream, past the recreation centre for the Royal Canadian Mounted Police which stands adjacent to Long Island Lock, for almost five miles to Black Rapids.

The long curved dam which can be seen by walking across the lock channel is over 700 feet in length and stands 30 feet high. It was the intent of the engineers to permit the water to crest the dam and flow down in a magnificent wide curtain like at Edmonds, Clowes and Black Rapids except this would have been about double the length of either of the other three. Unfortunately, this one dam displayed evidence of erosion even before it was completed and a by-pass channel had to be provided to carry off the excess water. Even this measure did not succeed entirely for in the summer of 1836 the dam washed out, destroying the use of this section of the canal. It was hurriedly reconstructed, this time with a firmer foundation, and the canal was in operation again within two months of the failure.

BURRITTS RAPIDS
Lock 17
Lift 9.0 feet
Kingston 84.0 miles

Adjacent to the roadway 4.8 miles north-west of Merrickville, 9 miles from the junction of this road and Hwy 16.

Before visiting the locks we suggest you stop at Burritts Rapids and become acquainted with the friendly people and the interesting history of this community. For instance, this is the site where the first bridge to cross the Rideau River was built. Financed by public subscription, which means the eight families living in the district paid for it, the bridge was opened in 1824, two years before the building of the canal system. The original foundation is still standing but the bridge itself has been replaced.

An interesting little fact is that the spring pickerel run in Burritts Rapids is equally as large as the more publicized run at Innisville, near Mississippi Lake which so many people view each year. Some houses in this village were constructed in the era of 1840, when more settlers began to arrive, and are in exceptionally fine condition.

Most of the property by the lock and up to the point of land is owned by the Dominion Government. Providing for a pleasant 15 minutes or half hours stroll on both sides of the bridge crossing the channel, the rich grassland by the canal makes a lovely cool carpet on which to play or relax. Some picnic tables are provided by the locks and a one night camping stand is permitted for boaters.

By Car...

CLOWES & NICHOLSON - LOCKS 18-19-20

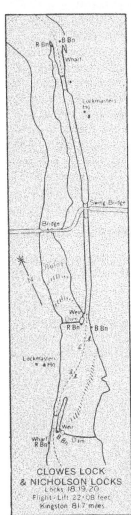

CLOWES LOCK
& NICHOLSON LOCKS
Locks 18.19.20
Flight-Lift 22·08 feet
Kingston 81.7 miles

Between Merrickville and Burritts Rapids. 2.2 miles north of Merrickville.

The entrance by road to Clowes Lock is marked by a small sign at the corner of a white post fence. Drive in towards the few buildings, then around the end of the house and down towards the lock. Please drive carefully and watch for children.

The arched dam by the lock is similar to others along the canal system and as usual presents a pretty sight. Two more dams lie a few hundred feet down the river and provide especially attractive scenes. Be sure to bring your camera. Parking space here is limited, so are picnic facilities. A pity!

Nicholson Locks are just ½ mile northwest of Clowes Lock as shown by the map. These are within easy walking distance or one can take the car north on the paved highway to the first main turning on the right hand side, then across the fixed bridge, and next the swing bridge, to Lock 19.

The approach channel to this lock is about ½ mile long and is cut through solid rock. A slightly wider channel leads down to Lock 18, a most remote and tranquil scene. Extensive and well tended grounds are provided and a pause along the banks to watch the frequent craft pass up and down is highly recommended.

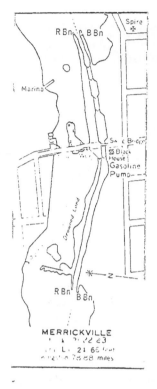

MERRICKVILLE

In Merrickville, 12 miles by road from Smiths Falls, 45 from Ottawa.

These three locks step down in a most pretty setting. The tree shrouded mainland and peninsula banking this man made navigation route is an ideal place in which to while away a few hours on a warm afternoon or evening. And many do, for everywhere along the Rideau Canal the attraction is irresistible and Merrickville is no exception. As usual, you will see the quiet fisherman lobbing out his lure to catch next morning's breakfast.

The blockhouse at the head of the locks is one of the two local defences still standing by the canal. The other and much smaller one is at Kingston Mills, the first stage out from Kingston.

The Merrickville blockhouse is a fine example of the superb workmanship of the stone masons and carpenters who fashioned this stalwart building. Constructed at the same time as the locks it was intended as a means of defending this section of the canal, at that time having an important road link with Brockville, from American forces should they attack inland. It is our good fortune no hostilities took place here for the blockhouse remains intact and in splendid condition. Inside is delightfully cool on a warm day. The bare stone walls are whitewashed and so is the ceiling with its massive wooden beams.

In this small quiet farming community are many old buildings some of which date back to the turn of the 18th century. The local folks will be only too glad to point them out to you and to explain some of the history of this area, including the founding of the first Boy Scout troop in Canada.

KILMARNOCK - LOCK 24

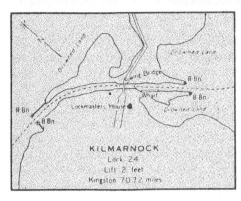

KILMARNOCK
Lock 24
Lift 2 feet
Kingston 70.72 miles

½ mile south off Hwy **43**, 5 miles east of Smiths Falls, 5.3 miles west of Merrickville.

Take side road west of long curve of Hwy **43** at above distances and follow for ½ mile to the lock station.

A single lock, in rolling farm country far away from traffic except for the pleasant passing of large and small craft. There are no particular distinguishing features at this lock but it is in a pretty spot and is one more link in the chain of interconnecting locks and lakes forming the Rideau Canal system.

EDMONDS - LOCK 25

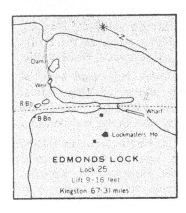

EDMONDS LOCK
Lock 25
Lift 9 · 16 feet
Kingston 67 · 31 miles

2½ miles east of Smiths Falls.

Take road at apex of the junction of Hwys 15 and 29 (Lombard and Brockville Sts.) in Smiths Falls for 2¼ miles then turn left for ¼ mile to the lock.

This lock is not accessible from Hwy 43 but one can drive down to the river's edge to within a few feet of the lovely arched dam that stands here. The lock can just be seen across the channel. Although limited in size some good picnic spots are to be enjoyed here.

The prime feature of this station is the lovely curved curtain of water draping the dam from end to end. Standing about six feet high, the dam is 8 feet wide at the top and over 250 feet long. It serves to back up the water to the foot of Old Sly Lock for a navigable channel.

OLD SLY - LOCKS 26-27

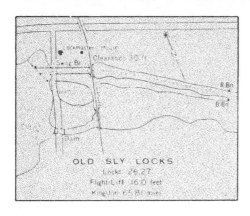

In Smiths Falls at Carthage St. off Hwy 43 (Queen St) at the eastern limits of town.

Flanked by a hefty road bridge and a not overly attractive railway bridge these two locks are a popular rendezvous for local residents. The small body of water a few hundred feet south of the locks looks natural enough with its attractive falls yet it was hand hewn through deep rock in order the water flow could be controlled.

The channel flowing east towards Ottawa follows an especially winding route before it straightens out at Edmonds Lock, some one and a half miles distant.

25

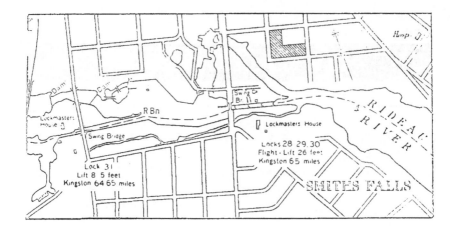

In Smiths Falls, 59 miles by road from Kingston, 48 from Ottawa.

Two separate sets of locks are located in Smiths Falls. The second set, Old Sly, was described in the previous section of this tour.

The Smiths Falls Locks lie adjacent to Hwy 15 and under the swing bridge at Beckwith St., the main thoroughfare of Smiths Falls. The Chamber of Commerce provides summer long tourist information from the booth in Municipal Park alongside the canal, and usually have in stock the official navigation charts of the Rideau Canal, for sale at fifty cents each.

A large docking basin is situated immediately by Beckwith St. The three locks leading to the east step down to a much larger basin where the Tops Motor Marina is so favourably situated.

Car parking is available on some of the side streets by the canal, on Beckwith St. and at the Tops Marina Hotel for the many who stop here for lunch or dinner or for overnight accommodation.

A most friendly town, Smiths Falls welcomes all travellers holidaying along the Rideau Waterway.

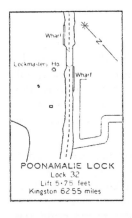

POONAMALIE LOCK
Lock 32
Lift 5·75 feet
Kingston 62.55 miles

1.7 miles west off Hwy 15 immediately south of Smiths Falls.

The side road is located just 1.3 miles south of Smiths Falls town limits and is marked by the usual green pointed sign. Follow this road to the lock, keeping to the right.

The final approach is through a beautiful stand of cedars offering a most pleasant aromatic flavour to the cooling air. Poonamalie, pronounced Poon-a-malie, was so named by the military engineer in charge of construction at this site because of the close resemblance its line of cedars bore to the characteristic shadows cast by screens of bamboo canes lining many of the roads in India. The correct Indian spelling is Poonahmalee; presumably the Canadian version is an error in dealing with an unaccustomed language.

The dam here controls the storage of water back through Big Rideau Lake to Narrows Lock, a distance of nineteen miles. By automatic regulation of the flow rate sufficient water can be kept in the canal system to Ottawa 61 miles away to permit safe navigation, yet a huge reserve can be held back for use during the busy locking months of July and August.

The Lockmaster's house stands opposite the head of the lock and dates back to the early days of the canal. A few quiet picnic spots are available on the well tended banks overlooking the lock.

1.5 miles south-west off Hwy 43 at Port Elmsley. Or turn off Hwy 15 at Lombardy and drive 7.7 miles through Rideau Ferry to just past the Drive-in Theatre near Port Elmsley.

Upon reaching the north end of the green painted bridge crossing the canal look on the east side for a sharp turn-off on to the narrow dirt road leading down to the channel. Please be watchful for children when driving down. The locks are 1/3 mile apart, one on each side of the bridge.

The story of this branch line of the Rideau Canal is an interesting one. When the Rideau Canal was completed in 1826 the people of Perth had no desire to be left high and dry from its commercial potential so they undertook an ambitious project to channel the River Tay along its path from Perth to near Rideau Ferry and build four locks, these to be of a much lesser standard and size than those constructed by Col. By.

The task was eventually completed in 1832 and the canal branch used for commercial navigation, even to the operation of a steamer service between Perth, Kingston and Bytown, now Ottawa. However, the toll revenues derived from the primary traffic of barges plying from Perth to the Rideau Lakes were insufficient to maintain the system in good condition and it gradually deteriorated. At this stage the federal government took a more active interest and constructed the two Beveridge Locks in 1886 to replace the worn out original locks.

Today this branch is used only by small pleasure craft, for Perth has long since abandoned this waterway as a commercial route and has even permitted two fixed bridges having only slightly more than a five foot clearance each to be built over the canal.

The lower of the two locks provides spacious grounds for picnics. The upper lock has limited parking facilities and recreation area but commands a handsome view up and down the channel.

NARROWS - LOCK 35

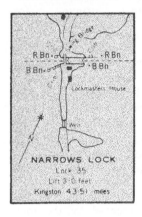

4 miles west off Hwy 15 at Crosby.

Crosby is 36 miles from Kingston and 22 miles from Smiths Falls. Enter Crosby on Hwy 42 for 1/10 mile only. Look on right for sign NARROWS LOCK RD on road leading between two historic but discarded stone buildings. Follow this road for 4 miles, keeping to the right.

Approaching the lock the water on the left is Upper Rideau Lake. On the right is Big Rideau Lake. From this point all water flows north east to Ottawa, 80 miles away, to empty into the Ottawa River then down to the St. Lawrence River to the Atlantic Ocean.

You may notice the waters on either side of the dam look almost level. Actually, there is a difference of only 3 feet. Originally these two lakes were one, joined together by a narrow channel where the lock now stands. To permit navigation this channel required to be deepened, quite a task when all labour was by hand and the lake bed of heavy rock. Instead of excavating the engineers decided to dam, raise the water level on the west side by a few feet, and install one lock thus saving onerous construction time and labour costs.

The road continues past the swing bridge to Westport and Perth through some beautiful backwoods country.

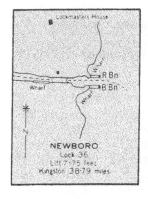

Lockmasters House

R Bn
Wharf
B Bn

NEWBORO
Lock 36
Lift 7.75 feet
Kingston 38.79 miles

4 miles off Hwy 15 at Crosby.

Follow Hwy 43 west of Crosby through Newboro to Stirling Lodge. Take side road in front of Lodge down to lakefront and lock. The main road continues on for 5 miles to Westport.

This single lock stands alone in a tree shrouded haven looking down to Newboro Lake and some of the many islands. One of the busiest locks in the system, the attendants here are kept hopping passing the many boats up and down between Upper Rideau Lake and Newboro Lake, a difference in level of 8 feet. At this lock the water flows south towards Kingston, 39 miles distant. At Narrows Lock, the next in the system towards Ottawa, all water flows north east to the Ottawa River at Ottawa.

Two privately operated trailer camps are located by the lock and conveniences are provided for campers. Boats may be rented from the numerous small docks lining the shore and Stirling Lodge has a large servicing dock and ramp for its guests.

The village of Newboro is small and has but a few stores. These nevertheless are well stocked with provisions and general supplies for your galley or picnic basket.

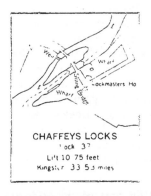

CHAFFEYS LOCKS
Lock 37
Lift 10.75 feet
Kingston 33.5.5 miles

5.2 miles west off Hwy 15 between Elgin and Crosby.

The turn-off point on Hwy 15 is 2.2 miles north of Elgin, 1.6 miles south of Crosby, and is well marked by signs. Follow this paved road through well wooded country to the parking area by the lock basin.

The approach wharf at the head of the lock has tie-up facilities for about 14 boats. The Opinicon Hotel offers a 400 foot dock for use of guests and Alford's Marina has an excellent servicing dock.

A first rate picnic area is provided under the trees alongside the upper approach wharf or stop in at the hotel for lunch or dinner and enjoy a few hours in this delightful spot.

The lock overlooks Lake Opinicon, renowned for its fishing pleasures and beautifully wooded shore line. In the opposite direction, up the channel past Dorothy's Lodge for about one mile, is Indian Lake after which lies Newboro Lake.

A singularly attractive part of the Rideau Waterway, the resort area of Chaffeys Lock has long enjoyed a reputation for the beauty of its location. In this remote and tranquil haven one cannot help but relax and laze in absolute peace. Should you have a desire for fishing there is a choice of twenty lakes all easily accessible without portaging. Well known as one of the two best areas in all the Rideaus you'll thrill constantly to the snap of the line and the joy of hauling in really big ones.

Should your larder need replenishing groceries and supplies may be purchased at the hotel commissariat or at the general store only a few hundred feet back from the water's edge.

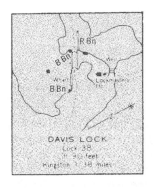

DAVIS LOCK
Lock 38
H 90 feet
Kingston 38 miles

5.7 miles west off Hwy 15 at Elgin.

This sideroad is not marked by a highway sign. The best information we can provide is to drive into Elgin to the corner where the Hotel Elgin stands. Turn west at this corner and proceed straight ahead, crossing Hwy 15. After about one mile this road divides. Take the right fork for Davis Lock. (The left leads to Jones Falls). At 2.7 miles from Hwy 15 this road divides again, a branch turning right to Chaffeys Locks. Keep left for Davis Lock just 3 miles ahead. You shouldn't, but if you do happen to become lost, don't worry. These back roads are only a few miles long and if you keep heading west you'll shortly arrive at a lock station.

Long regarded as an exceptionally beautiful part of the Rideau Waterway, Davis Lock is enjoyed by many who come to picnic by the shores of this lovely bay. Fortunate are those who own cottages hidden amongst the trees. To the north-west the channel leads into Lake Opinicon and Chaffeys Lock; to the south-east it swings into Sand Lake and down to Jones Falls.

The fishing in all this district is superb. We have never yet heard of a sportsman who doesn't sigh when it comes time to leave for home. Be sure to bring your rod and a good sized creel when you holiday along the Rideau Waterway.

JONES FALLS - LOCKS 39-40-41-42

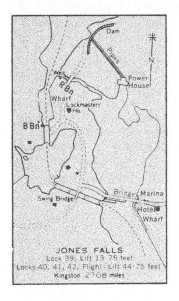

JONES FALLS
Lock 39; Lift 13.75 feet
Locks 40, 41, 42; Flight; Lift 44.75 feet
Kingston 27.08 miles

2 miles west off Hwy 15 just north of Morton.

This paved side road is well marked by signs but slow down as the turn off the highway is deceptive. Follow this road for 2 miles to Hotel Kenney, across the first bridge, up and over the swing bridge then right to a parking area.

From alongside Lock 40 there is a commanding view down to the basin leading from Kingston. Many say the thrill of seeing this scene for the first time is without equal. The locks at this station are of standard size, 134 feet long by 33 feet wide but are immense in depth since they provide the greatest average lift in the entire Rideau Canal system. From the basin to this point the three locks lift up 44.75 feet and Lock 39, on the far side of the upper basin, lifts a further 13.75 feet.

To the left of Lock 39 can be seen the original blacksmiths shop built in 1827 to forge the ironwork for these locks. Shame to our government for neglecting this historic building, one of the few remaining stone buildings constructed at the same time as the canal.

The beautiful arched dam cannot be seen from this point since it lies behind the hill to the right of Lock 39. One can walk, or drive, back past Hotel Kenney to turn left at Pine Hill Lodge, then left again after 0.2 miles at the two mail boxes. Keeping to the right, follow the road to the dam, a total of 0.6 miles from Hotel Kenney.

Walk over the dam to the hill overlooking Lock 39. On the way you will cross a small wooden bridge. Step down the few cement stairs to see on the right the rotted tips of two foundation pillars placed in 1827 to dam back the water churning down this rocky gorge.

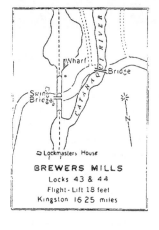

BREWERS MILLS
Locks 43 & 44
Flight - Lift 18 feet
Kingston 16.25 miles

1/3 mile west off Hwy 15 between Joyceville and Seeleys Bay.

A sign post on Hwy 15 reads BREWERS MILLS RD and marks the side road which is 3.2 miles north of Joyceville and 13 miles south of Morton.

These locks are sometimes referred to as Brewers Mills, a name naturally attributed to this site where a Mr. Brewers operated one of two mills, the second being down at Lower Brewers, which also can be confusing since it is often called Washburn. We say this only in the event you are referred to Brewers Mills or Washburn.

In this peaceful glen one can almost hear the roar of traffic on Hwy 401 a dozen miles away. Certainly the quietness that prevails in this lovely corner of Ontario gives no indication of the terrors of the rampant swamp fever that struck down the builders of this canal. This fever, a severe form of tropical malaria caused undoubtedly by the hot and humid conditions aggravated by the decomposing of vast stretches of deep slimy marshes, was not confined to the immediate region of the Rideau Canal. Well documented reports show that in 1828, two years after the Rideau Canal project was started, almost all the populated areas of Upper Canada were in the throes of this pestilence.

But these difficulties are past and now we can enjoy the efforts of those who wisely and well did complete this fascinating canal system, of which Upper Brewers is one of the many lovely stations.

LOWER BREWERS - LOCK 45

WASHBURN
Lock 45
Lift 13 feet
Kingston 14·47 miles

1/10 mile west off Hwy 15 between Joyceville and Seeleys Bay.

The side road is just 1.2 miles north of Joyceville and is marked by a sign.

Often referred to as Washburn, this lock is so close to the highway it could readily be seen were it not for a rise of ground between the two. Drive the few feet to the lock and the highway will soon be forgotten.

From a select vantage point one can see at this station three simultaneous functions of the Rideau Canal. First the locking of boats, second the damming of rapids, third the harnessing of water power to drive turbines for generating electricity. Only four lock stations make use of this last feature and all except two (not counting Beveridges which is not part of the original Rideau Canal system) have dams. The large flume stretching from the dam to the power house carries water down to drive the turbine.

A natural rock bound pond, reminiscent of the old time swimming hole, is a favourite spot for the youngsters on a hot day. The many pine trees provide ample shade and screening for a quiet picnic in this picturesque setting.

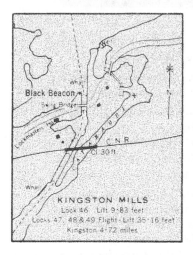

KINGSTON MILLS
Lock 46   Lift 9·83 feet
Locks 47, 48 & 49 Flight·Lift 35·16 feet
Kingston 4·72 miles

1 mile west off Hwy 15 at a point one mile north of Hwy 401.

There is no highway marker to identify this side road leading to the lock station. However, if one is travelling out from Kingston, or off Hwy 401, two historic site plaques may be seen off the right shoulder of Hwy 15. The first is one half mile from Hwy 401 and the second one mile. Turn sharp left at this last marker opposite Farrars Grocery store. Mr. Farrar also serves those who approach from the north for his establishment provides a good landmark as well as plentiful supplies of provisions for campers.

Drive down the mile long road, noting there is a camping and trailer park site on the right, and over a small bridge to a parking area which actually is the top of a dam used to hold back the water for 10 miles up to Lower Brewers. The comparatively large body of water here is Cataraqui River, flowing down from Dog Lake and Jones Falls. Some of it passes through the locks but a large flow is permitted down the flume near the small bridge to the hydro-electric power plant below.

Lock 49 at the base of the flight of three locks provides the first lift up from Kingston some four and a half miles away. From this point a series of 14 locks lift craft up 162 feet to the summit of the Rideau Canal system at Upper Rideau Lake. The three locks here provide an impressive sight in the steeply banked gorge as they seem literally to gulp vessels up to the docking basin and the fourth lock. The railway bridge that passes overhead is the main line of the Canadian National Railways from Montreal to Toronto.

The plateau by the small docking basin is one more of the many popular picnic areas along the Rideau. Here under the two towering maple trees one can take in a pleasing scene, changing each time the lock gates open.

Holiday Haunts

A selection of popular vacation centres from which to enjoy The Rideau Waterway.

KINGSTON

Popularly known as the "Limestone City" Kingston is the second largest community in Eastern Ontario. Steeped in the history of North America much evidence remains today testifying to its early military importance. The community is rightfully proud to have fostered Queen's University, founded by Royal Charter under Queen Victoria in 1841, and the Royal Military College, Canada's equivalent of West Point in the U.S.A. and Sandhurst Military College in Britain.

Kingston has much to offer the transient guest. A high standard of accommodation is provided by the 26 motels and hotels located within the immediate environs and rooms are sensibly priced. Many fine restaurants and dining rooms have earned for Kingston an enviable reputation for good food. Two outstanding summer attractions always popular with visitors are the 80 mile boat cruises through the beautiful 1000 Islands aboard the luxurious Miss Kingston and Lady Kingston; and Old Fort Henry where colourful displays of precision marching by stalwart guardsmen are presented daily. Be sure to visit the Royal Military College, Fort Frederick, Murney Tower and Martello Tower, the latter containing four floors of exhibits.

Many fine shops are located downtown and at a shopping centre mid-way along Princess St. The Kingston Yacht Club is handsomely located at the southern edge of the city looking across to Fort Frederick and Fort Henry. Kingston provides top rate marina and docking services and is a favourite stopping point for boats about to enter or leave this end of the Rideau Canal.

BATTERSEA

Adjacent to Loughborough Lake, in heart of Battersea Lakes. Take Interchange 103 off Hwy 401 and follow the good road 15 miles to Battersea. Or 9 miles from Seeleys Bay off Hwy 15 on narrow, winding but pretty road. 13 miles from Jones Falls on good gravel road, 25 miles from 1000 Islands Bridge, 91 from Ottawa. A peaceful haven from the hustle and bustle of hot cities. No traffic or parking problems here. Famed for fishing and guide service. Loughborough Lake is 21 miles long studded with 114 well treed islands and teeming with bass, northern pike and lake trout. Fresh bait stocked. Many other excellent lakes nearby including Dog Lake and Cranberry Lake. Reputed best guide service in all Ontario. Boats and motors rented. 2 hotels, cabins, cottages and private homes provide plentiful accommodation. Good shopping facilities in this very friendly town. For naturalists an abundance of wildflowers including the beautiful trillium plant can be found along every tree canopied trail.

Local gossip says you're bound to catch the Battersea Disease after just one visit, then you'll be back year after year. A well known resident, Marguerite Stoness provides own voluntary service to supply information on accommodation, fishing forecast, guides, etc. See her for your fishing license. A one woman public relations counsellor and publicity agent Marguerite can be telephoned at Inverary 68 Ring 13 (area code 613) or ask for her when you reach Battersea.

SEELEYS BAY

Just ¼ mile off Hwy 15, 23 miles from Kingston, 35 from Smiths Falls, 83 from Ottawa. Or follow Hwy 32 from Gananoque total 14 miles or from 1000 Islands Bridge 23 miles. A progressive community that goes out of its way to help make your vacation a complete success. Well stocked stores provide fresh and canned foods, hardware and marine supplies, etc. There is a barber shop, coin laundry, post office, a good restaurant with ice cream bar, four churches and a doctor and a marina. The Federal Government operates a 400 foot dock serving marine travellers. Two motels, 1 hotel, 4 lodges and numerous cottages and cabin establishments are located within or in the immediate vicinity of Seeleys Bay. Some tenting and trailer facilities with complete conveniences are available.

Fishing licenses, fresh bait, lures and accessories for rod and reel are sold here. Experienced guides may be hired by the day or week to lead you through the miles of beautiful lakes in the district. Many folks settle here each year for their holidays. Dad can fish to his heart's content and the remainder of the family can laze on an excellent beach, take a boat tour of the 1000 Islands or do some shopping in nearby Kingston or other neighboring towns. Side trips can be made in a matter of minutes to Jones Falls, Brewers Mills or Washburn Locks to watch the always fascinating sight of holiday boats locking up and down the Rideau Canal.

A sign on Hwy 15 immediately north of Morton reads "Jones Falls 2 Miles". This is the only invitation you need to enter a part of the Rideau Waterway many consider to be the most scenic spot in all North America.

Come see for yourself the flight of four locks rising almost 59 feet in a great sweep that leaves you breathless. Stand at the head of Lock 40 under the towering maple trees and gaze into the chasm of this massive construction of hand hewn stone blocks that have weathered time these last 134 years. Look out over Lock 41, across which the swing bridge stretches, down to the bay where the Kenney Hotel can just be seen through the dense foliage of the beautiful trees. A picture unto itself it has to be seen to be appreciated. The tranquility of the Kenney Hotel nestling in the bay at the bottom of the locks with its own private swimming pool reflecting the icy blue water, the greeness of the trees, the splashing of the sluices, the boats locking up and down, the brute weight of the lock gates with their trim of white provide a scene you will long remember.

Stand at the edge of the basin at Lock 40 and look over to the old blacksmith's shop and Lock 39 with its massive entrance wings. Walk around to the great arched dam and marvel at this engineering achievement, built entirely by hand.

The beauty of Jones Falls is incomparable.

Holiday Haunts...

CHAFFEYS LOCKS

The beautiful tree lined channel at the head of Chaffey's Lock is just about the prettiest spot along the entire Rideau Waterway. With a fresh, cooling breeze fanning across the water a more pleasant place on a summer's day would indeed be hard to find. Those who built lodges and cottages and a marina amidst this haven of trees made an enviable choice. Those who vacation here are favoured with the best nature and friendly service can provide. One can sit by the hour alongside the channel watching the colourful boats cruise up and down and talking to the passengers as they wait to lock through. Or take a boat yourself and head up channel under a canopy of foliage into Indian Lake a mere mile away or lock down into Opinicon Lake and tour its fascinating shore line to beautiful Davis Lock.

The accommodation provided at Chaffey's is superb. With an excellent hotel and lodges and cottages, a good general store and a marina there is all one could ask for. The fishing here has a reputation going back to the days of the Indians and is as good now as ever. Guides and boats may be hired and, if that big one did get get away, you can sit on the Liars Bench at the Opinicon and tell folks all about it. You should be warned though, the locals have likely seen bigger ones than you can imagine.

Look for Chaffey's on our highway map, off Hwy 15 between Elgin and Crosby.

43

Located at the extreme western tip of Upper Rideau Lake, at the end of paved Hwy 42 just 9 miles from Crosby at Hwy 15, and only 45 miles from the 1000 Islands Bridge, Westport commands a particularly fine vantage point over this section of the Rideau Waterway. A small peaceful village of some 700 persons it is especially popular with vacationers and is a thriving community during the summer months. Many families have built their cottages here along the quiet and well wooded shores of the bay and some excellent estate areas in the immediate vicinity offer cottage lots, and cottages too, for those who wisely select this region for their annual holidays.

The Upper Rideau Lake provides the highest point of navigation in the Rideau Canal system. At this point the watershed divides to flow south to Kingston and north-east to Ottawa. From Westport one can boat through 25 miles of scenic waters, locking only once at Narrows Lock, to past Rideau Ferry almost to Smiths Falls. Some forty lakes lie within a 15 mile radius of Westport and every one provides top rate fishing. All can be reached by the half dozen roads that radiate outwards from this popular centre. Interesting tours can be made via these back roads to Perth, or to Loughborough Lake and around to Battersea and Jones Falls. Having some excellent hotels, lodges and cottages, a government dock with free tie up facilities and a fish hatchery, the region offers vacationers carefree days of rest and relaxation.

PORTLAND

Often referred to as Por ₁and-on-the-Rideau, this village stands on a bay of Big Rideau Lake which is nearly 20 miles long and has over 300 islands. In a most favourable location for the transient pleasure craft or for those who wish to launch their boat for a few hours sailing this pleasant community enjoys a summer long popularity year after year.

Located alongside Hwy 15, Portland is so laid out that once you enter it you quickly forget the highway and the swish of the traffic. Being only 16 miles south of Smiths Falls, 44 from Kingston, 31 from Gananoque and 42 from Brockville via Hwy 42 it is convenient to many centres in this vacation area.

Accommodation within Portland is confined to one of the two marinas that have cottage facilities. Both marinas incidentally have complete facilities for effecting repairs to boats or engines. A good half dozen cottage establishments are located within the district and there is a motel and a cottage resort a few miles north along Hwy 15 at Otter Lake.

The people of Portland are most friendly and helpful and can be depended upon to do their best to make your holiday a pleasant and enjoyable one.

For those who may be considering the erection of a cottage or boat house there is an excellent lumber yard, staffed by people who know the business, where you can see all the latest items and get the best hints the experts can give.

Standing on a lovely winding country road that extends from Lombardy, on Hwy 15 to Perth some 12 miles distant, Rideau Ferry is a well known location to those who frequent the Rideau district for water sports and real country living. Known originally as Olivers Ferry, this small village has seen the settling of the Rideau Lakes district from the early 1800s, yet itself has remained almost untouched by the passing stream of immigrants and commerce. In the early days the direct route from Brockville to Perth necessitated passage over Rideau Lake and it was at this point a ferry service provided access across the quiet waters. In a sense this service is still available for a narrow swing bridge straddling this •comparatively narrow part of Rideau Lake allows ready passage for motor vehicles.

Today Rideau Ferry is a mecca for the boating fraternity. Launching and docking facilities are available and boats may be rented. Each August boaters come from up and down the lakes to take part in the Rideau Ferry Regatta, an impressive water festival that increases in popularity each year. Three well equipped marinas serve the many craft that pass here and a typical country store provides for groceries and odds and ends.

In Rideau Ferry is a lodge-motel, an inn and many cottages. Within 5 minutes drive is Bass Lake, a beautifully secluded spot with a well-known lodge.

Be sure to include Rideau Ferry in your tour of the Waterway.

PERTH

The oldest community in the Rideau Lakes district, Perth is today a charming little town with a most winning personality. Known far and wide for the friendliness of its residents it will provide you with an enjoyable afternoon or evening in Stewart Park, or strolling Drummond and other tree lined streets to admire houses dating back over a hundred years.

Conveniently located at the junction of Hwys 7 and 43, and only a few miles from Rideau Ferry and Hwy 15, Perth is the community town for the many vacationers who flock annually to the numerous resorts and cottages encircling the few dozen lakes in the immediate district. Settled by Scots in 1816 the town exhibits today much evidence of its link with the past. The Town Hall was built in 1862, 5 years before Confederation; the clock in the tower has chimed each quarter hour for more than 90 years; there is at least one house dating back to 1816, still lived in and more comfortable today than many a modern structure. The Scottish architectural style can be readily identified in many buildings lining downtown Gore St., though modern facades distract from the clean simple lines of these well constructed habitations.

Well known for its small but pleasant stores and business establishments, Perth caters especially to the needs of vacationers. Located on the banks of the Tay River, that flows from Christie Lake to Rideau River near Rideau Ferry, it warrants a visit by every touring family.

SMITHS FALLS

The third largest community along the Rideau Waterway, Smiths Falls is favourably located at the junction of Hwys 15, 29 and 43 almost mid-way between Ottawa and Kingston. It provides the modern traveller with excellent accommodation and shops and is a most convenient place from which to start ones activities in this great vacation area. With two sets of locks, Smiths Falls and Old Slys, located virtually within the centre of town, this has long been a favourite spot for many to come and picnic by the Municipal Park and watch the boats sail through this important part of the Rideau Canal.

The 10,000 residents of Smiths Falls are intensely proud of their community, and rightly so. An active town healthily engaged in fostering sports activities for youths and in attracting industry to its environs, it presents an exceptionally clean and slightly dignified appearance that reflects the manner in which visitors are made to feel welcome.

Beckwith St. is the main thoroughfare and provides for the broadest main street in any Eastern Ontario town. With adequate parking facilities on both sides it adds to the pleasure of visiting the many shops along its route. For boaters there are docking facilities at the municipal tourist park within a moments walk to the Beckwith St. restaurants and stores. A supervised beach, swimming pool and pleasantly cooling shade around recreation areas is to be found within this park. The Hershey Chocolate plant on the eastern suburbs offers conducted tours each weekday.

MANOTICK

Located on the Rideau River some 10 miles south of Ottawa, the village of Manotick is rapidly gaining an important role in catering to visitors whether they come by car or on the many fine craft which ply their way this far along the Rideau Canal. Many cottages and modern homes line the river in this rolling farming country where a good proportion of the residents commute daily to their work in Ottawa. From one of the two bridges one can see Watson's Mill, an historic landmark that has been fully restored by the National Capital Commission and is open for visitors during the summer months, and the dam holding back the water for the canal.

An ambitiously planned community is being developed just south of Manotick and includes individually designed housing, a handsome golf course and an excellent marina with extensive docking facilities. Many boaters tie up here to avail themselves of the excellent services provided by the Carleton Golf and Yacht Club.

Watson's Mill, at Manotick

Ottawa, the Capital of the Dominion of Canada. Standing majestically at the junction of the three rivers which fostered its early development. Settled in 1826 by Lt. Col. John By during the building of the Rideau Canal, hewn from a wilderness of swamps and forests, growing with the thriving lumber industry, selected in 1857 as the Capital of the newly united provinces of Lower and Upper Canada and 10 years later, on July 1, 1867, of the Dominion of Canada. Becoming a government centre yet for decades almost forgotten by the government, mushrooming suddenly in the nineteen fifties and sixties to emerge as a truly beautiful city, one all Canadians and Americans should visit for a few days.

Crowning the heights overlooking the mighty Ottawa River, where passed Etienne Brule, Nicholas de Vignau and Samuel de Champlain and by which route was the great western half of the United States and Canada first explored, stands the beautiful Gothic style Parliament Buildings. The Peace Tower rises handsomely as a symbol of our freedom, a fitting tribute to the 66,650 Canadians who gave of their lives in the Great War 1914-18. On the beautiful lawns of Parliament Hill military displays of pomp and splendour thrill the visitor with a panorama of colour and sound.

A city through which entwines the Rideau Canal and the Rideau River, having a lake and over 10,000 acres of parks and playgrounds, a historic dam and 12 locks, many miles of scenic driveways, a shoppers mall and boat cruises through the very centre of downtown, Ottawa has much to interest the visitor. A well rounded sightseeing itinerary would include the Royal Canadian Mint, the National Gallery, the National Museum, the grounds of the residence of the Governor General of Canada, the Royal Canadian Mounted Police training centre, the National Aviation Museum, Bytown Museum, Ottawa City Hall, Nepean Point overlooking the very start of the Rideau Canal system, and many other equally interesting places.

Cruising—Kingston to Ottawa

The ultimate pleasure of the Rideau Waterway is to cruise its beautiful lakes, rivers and canals. No other system on the North American continent, and there are many exceptionally fine cruising routes in other parts of Canada and the United States, can match the perfect blending of this water route.

With twenty two locking stations having forty seven locks in the main canal, winding for 123 miles from historic Kingston along two rivers and twelve lakes each with its own individual charm, passing through the towns of Smiths Falls and Merrickville and finally through Ottawa, our Nation's Capital, where the last flight down to the Ottawa River provides a fascinating and fitting climax to an incomparable voyage, the Rideau Waterway never fails to thrill those who crave for an exciting boating tour.

Starting from Kingston, we'll follow this route right through to Ottawa. It is recommended you provide yourself with a good topography map or the navigation charts listed on page 70 of this guide. Information on marinas, docks and accommodation facilities are provided on page 71. Please remember, keep to the well marked main channel and do not divert from it, especially where the channel is narrow, as many weed beds can quickly foul your propellor.

The Rideau Canal at the southern end has its official beginning at the LaSalle Causeway, the bridge crossing the harbour at Kingston. The Cataraqui River finishes here after flowing down from Cranberry Lake. To the right is Old Fort Henry, The Royal Military College, HMCS Stadaconna a naval establishment, and on the left lies Kingston. An excellent marina is located on the bay just after the Causeway and three first class motels are on Hwy 15 near the marina.

The first set of locks are located at KINGSTON MILLS, almost 5 miles upriver. These are located in a rocky gorge, making an impressive sight. Note the first lock you reach is numbered 49. Altogether there are 47 locks on the Rideau Canal and 2 on the Tay Canal, a diversion to Perth. The locks are numbered 1 to 49, commencing at Ottawa. Observe also the boys and men fishing by the locks. Their counterparts will be seen through the whole system, right into downtown Ottawa. A small docking basin is provided just before the upper lock. Ablution facilities are provided here

Cruising · · ·.
and there is a small roadside stand just beyond the east side of
the station. An original blockhouse can be seen near the small
bridge to the right of the upper lock. Today it serves as a habitation!

Continuing up the Cataraqui River for almost 10 miles we reach
WASHBURN or LOWER BREWERS MILLS, a lovely peaceful spot.
Would you believe Hwy 15 is only 1/10 mile away? Next comes
UPPER BREWERS, a mere 2 miles along a winding channel. What
is now Cranberry Lake was once a slimy, thick fetid bog that spawn-
ed a fever killing many of the workers who built this canal. It was
necessary to flood this bog in order to smother its odorous properties, thus a spillway dam was constructed here and two locks
were placed to allow boat passage. A half mile after these locks,
where the channel opens into the lake, is a rocky promontory showing
the profile of a human face. This is named the Duke's Profile in
honour of the Duke of Wellington who did much to promote the
construction of this canal.

Up now through Cranberry Lake, one of the first man made lakes
in North America, and Little Cranberry Lake off which Seeleys Bay
nestles quietly offering many services to boaters. On again through
Whitefish Lake to the most attractive station of JONES FALLS.
We highly recommend you tie up here for a few hours, having lunch
or dinner on shore, and take a walk around the locks and the great
dam which lies over near the high television tower. Be sure to take
your camera with colour film on this jaunt. We have described Jones
Falls in other chapters of this guide so we'll not be repetitious
except to say again, it's a perfect haven.

Leaving Jones Falls, we are now 27 miles from Kingston and
nave climbed 134½ feet. Cruising through the beautiful channel
into Sand Lake we come to DAVIS LOCK, a pretty and well tended
little station. The placid waters along this part of the canal may
induce an hours fishing as you float in the many peacefully remote
bays. Up again and we enter Opinicon Lake, another fisherman's
paradise. Here is located a fish hatchery for use of Queen's University students, and two fish sanctuaries operated by the Ontario
Department of Lands and Forests. Fishing in these sanctuaries,
one in the southwest and one in the southeast corners of the lake
is strictly prohibited. Otherwise, enjoy yourselves in this beautifully
islanded lake.

Following the channel again through Lake Opinicon we come to
CHAFFEYS LOCK. Lifting up the one lock we enter a basin and

53

Entering Lock 39, Jones Falls

The quiet haven of Chaffeys Locks

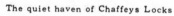

are presented with a picture of rare beauty. In this tranquil chamber
of pines and maples with a spot of blue sky overhead one cannot
resist the urge to stay and visit the almost hidden hotel and village.
After a wonderful meal and a saunter along the tree lined shores
in this most remote spot we climb aboard and slowly wind up the
channel to Indian Lake. Here we can continue to follow the channel
or we can turn to port and proceed past Dunn Pt., Benson Pt. and
Pollywog Lake (an intriguing name) to enter into Newboro Lake,
one of the most attractive lakes in the waterway. A sidetrip around
some of these islands and you will better appreciate the loveliness
of this country.

At the upper end of the lake lies NEWBORO LOCK providing
the final lift up before we come to Ottawa. Some tie up points are
provided by the shore before you enter the lock. Newboro Village
is just a short walk up past Stirling Lodge should you need provisions or a post office. A camping site with ablution facilities
lies adjacent to the lock. Many boaters camp here for the night as
the view in the late evening and early morning is one of the chief
reasons people camp out. Coming out of this lock we have risen
162 feet through 14 locks to a point 408 feet above sea level and
have travelled 39 miles from Kingston. We are now at the highest
level of the Rideau Canal system and from here on will be stepping
down to Ottawa. Note that from here to Ottawa the canal is marked
with the red buoys to port and the black buoys to starboard, the
opposite of the Kingston to Newboro section.

Coming into Upper Rideau Lake we again have a choice of a
diversion to the village of Westport at the extreme west end of the
lake. NARROWS LOCK lies to the east and drops only 3 feet into
Big Rideau Lake. We now travel over 19 miles to the next lock at
Poonamalie. A handsome large lake with over three hundred islands,
Big Rideau offers a side route to Portland where two marinas, one
with cottages, are located.

At the end of Big Rideau Lake is Rideau Ferry, a charming
little community. Here are located three marinas, this being a
popular centre for the boating fraternity. Passing under, or through
the swing bridge, we will see on the left the entrance to BEVERIDGE
LOCKS that leads to Perth, about 7 miles away. The channel leading up to Perth is reminiscent of the German and French canal
routes, and winds peacefully through remote country. However,
there is a severe limiting condition to travelling this quaint channel

and that is the low clearance of just 5 feet under two bridges in Perth. You have to duck pretty low to creep under these but Perth is a lovely town to visit and if you have the time and can manage the clearance you'll find this is a most pleasant trip.

On again through Lower Rideau Lake and we come to POONAMALIE, another attractive station and with a most unusual name for Canadians to use. On a quiet evening the pungent aroma from the cedars on either side fills the air almost to the end of the channel. This station incidentally controls the water flow rate throughout this part of the system to Ottawa. Past Poonamalie and we can see immediately ahead Smiths Falls with its two stations of six locks.

The SMITHS FALLS LOCKS are detached. Passing under a railway bridge we enter Lock 31 and emerge in a basin where tie up facilities are provided. The next three locks of this station are past the low lying swing bridge and these lead down to a bay where there is a marina hotel. As it takes up to one hour to lock through these three locks many passengers take the opportunity to walk the few moments to the shopping district just a few hundred feet away. This is a good opportunity also to purchase Navigation Charts from the Chamber of Commerce information booth or main office if you have not already procured them. Ask anyone where these are located.

We are now slightly more than half way from Kingston, having ahead of us 12 stations with 27 locks to the Ottawa River. The route from here on passes through low lying countryside, the more rugged hills of the Precambrian Shield having been left behind at Rideau Ferry.

OLD SLY LOCKS lie at the opposite end of the bay from Smiths Falls Locks. Passage through here is reasonably quick and we enter a winding channel that demands navigational care before reaching EDMONDS LOCK, a quiet single lock with a beautiful long arched dam. KILMARNOCK LOCK lies 3½ miles further along through a fairly broad part of the river, almost a lake, but keep to the channel as the water is very shallow. Kilmarnock is as pretty as its namesake in Scotland. Though it may seem incongruous to have such a large lock to effect a change in level of only two feet, the lock is of standard size in order there be no bottleneck in the system that would thwart the passage of military traffic for which this canal was designed.

Poonamalie, near Smiths Falls

Old Sly Locks, Smiths Falls

57

On now for another 8 miles through lovely farming country to MERRICKVILLE, once an important milling community and now a quiet rural town peacefully enjoying this corner of the world. The large blockhouse to the right is in well preserved condition and warrants a visit by those interested in a touch of history. The three locks here drop almost 25 feet to a lovely bay to the left of which is a large marina.

Another 2½ miles further on lies CLOWES LOCK and immediately after that NICHOLSON LOCKS. One could almost count these three locks as one station. Each of these locks are popular with camera fans and are indeed a pretty sight in mid summer.

Past a few more miles of lush green fields and we approach BURRITTS RAPIDS, one more of the almost countless lovely rural villages alongside or near the route of the Rideau Canal. Enjoy the beauty of this countryside as you travel, untouched as yet by commercialism.

We now have almost twenty five miles of sailing before reaching the next set of locks at Long Island, just this side of Ottawa. The Rideau River is relatively narrow but the channel is safe and easy to follow. Two marinas and a camping area are located on the west side of this run towards Manotick.

The LONG ISLAND LOCKS provide a haven for Ottawans on summer evenings and week-ends. This means of course the usual attendant of speed hounds must display their talents, or at least evidence of their exuberance, so beware the accompanying waves and water skiers. The locks step down twenty five feet. We are now almost 152 feet below Upper Rideau Lake and 125 feet above the Ottawa River just 14 miles away.

Coming out of Long Island Locks we pass the Recreation Grounds for the Royal Canadian Mounted Police on the right. Proceeding down river through pleasant countryside, with Hwy 16 on the left, and past another large marina, we cruise the 5 miles to BLACK RAPIDS LOCK. The one lock here is sometimes quite busy with local traffic but any wait is never a long one and we are soon on our way to Ottawa.

Ottawa Uplands Airport, a large civil and military terminal, lies just over the right bank and for this reason some flights passing over are at a low altitude. The entrance into Ottawa is a gradual change from rural countryside to a few suburban estates. Ottawa has been so well planned with respect to the parkways parallelling the canal that one can cruise through the centre of downtown and yet hardly be aware of the traffic and the three hundred thousand people encircling the final stage of this historic route.

HOGS BACK is situated at the end of Mooneys Bay, a popular summertime playground for this Capital City. The swing bridge here remains closed during peak traffic periods so best check your official timetable to safeguard having to wait an extra half hour or so. Not that it will be wasted as there is much to see and a fine marina will look after your needs. A set of rapids lie just over the dam to the right and will provide you with a good picture or two. This is the Rideau River which separates from the canal to wind its own way to the Ottawa River, just 5 miles away.

Proceeding on to HARTWELL LOCKS, one mile down, with Carleton University to the right and with some intriguing inlets on the left in which are offered glimpses of private residences, one may be excused for not recognizing we are now well within the environs of this city. Past Hartwells and we come to beautiful Dows Lake at the end of which is another marina. Each year, in the last two weeks of May, Ottawa presents the Canadian Tulip Festival, a panorama of colour and beauty that sweeps along the driveways by Dows Lake and by the canal in a series of fascinating displays.

Continuing past Dows Lake we enter what many consider to be the finest part of the voyage, the cruise through downtown Ottawa. Passing under the first bridge from Dows Lake the well treed banks on either side rise high to almost envelop us in a canopy of foliage. Cruising slowly we emerge to see an inlet on the left, quiet and serene in this most natural setting. Under a bridge and next on the left behind the row of poplars are the Fair Grounds, then another narrow inlet may be seen through the arch of a small road bridge. We are now passing through a residential district close to the centre of the city. On the right is part of the University of Ottawa. Next under, or through, a lift bridge then under a massive million dollar high rise bridge, a link in Ottawa's new Queensway freeway. We are approaching a sharp turn to port after which on the right are the railway lines leading to Union Station three quarters mile ahead. A new terminal station is now under construction in the south-eastern suburbs of Ottawa and when completed in 1967 these tracks will be removed. Passing under two more bridges, the Laurier and

Cruising...

the MacKenzie King respectively, we have reached downtown Ottawa. On the right 'stands Union Station, and the Chateau Laurier Hotel can just be seen behind the massive stone bridge towering ahead of us. On the immediate left the federal government are constructing a Performing Arts Centre, planned to be opened in 1968. Beyond the high bank on the left, on a slight rise of ground and in a most impressive setting overlooking Ottawa to the front and the historical Ottawa River to the rear, stand The Houses of Parliament, the seat of Canada's Government. We will be able to see these beautiful Gothic style buildings as we sail out of the last lock to the Ottawa River.

The Rideau Canal winding through Ottawa

Enshrouded by the stone arch of the bridge we approach the last high point of the Rideau Canal. Ahead lies the OTTAWA LOCKS, or the Giant's Staircase — a name most appropriate, down which we will tread 79 feet to the Ottawa River below. Entering the first lock and being greeted by the friendly attendants we have an opportunity to look about us. On the immediate right is the Chateau Laurier Hotel, built in 1912 and recently completely renovated, then ahead on the right is a railway line leading over the Alexandra Bridge, erected in 1901 and more commonly referred to as the Interprovincial Bridge since it links the provinces of Ontario and Quebec. Ahead is the Ottawa River and Hull, settled in 1800 by Philemon Wright who came from Concord, Mass. and who is credited in some circles with suggesting the Rideau Canal be built. On the left, half way down the flight of locks, stands a lonely stone building. This was the headquarters of Col. John By and here was conceived many of the plans for the design and construction of the Rideau Canal stations through which we have travelled from Kingston. You cannot come this far without tying up at the bottom of the locks and walking back to this historic site, now a museum, and indulging in the intimacy afforded by the many collections.

Sailing out from the lower lock we have completed our 123 mile long voyage, but there is still a treat or two left. Proceed out to mid-river then look back to the majestic Parliament Buildings with the Peace Tower rising high at the centre and the magnificent Parliamentary Library with its beautiful cupola. The pulpwood logs you can see in the booms or stored high on the shore by Hull have been floated down from north of Val D'Or and Senneterre, some 200 miles up the Gatineau River. Sail down the Ottawa River for one half mile to where, on the right, the Rideau River falls 35 feet into the Ottawa River in two lovely falls. Samuel de Champlain made particular mention of these in his diary. Behind the falls stands Ottawa City Hall, and to the left is the French Embassy and further left again the official residence of the Prime Minister of Canada, a grey stone building standing on a point of land.

Many boaters choose to continue down the Ottawa River to return to the St. Lawrence River by The Lake of Two Mountains, near Montreal. The majority however, prefer to return by the Rideau Canal to see again their favourite localities and to search out points of interest missed on the way up. Whichever way you go, Bon Voyage and Happy Holidays.

The Rideau Falls and Ottawa City Hall

The Rideau River at Hog's Back, Ottawa.

The maintaining of an adequate flow of water through the Rideau Canal system is an important function of the federal Department of Transport who are responsible for the operation of the canal.

Fortunately, the Rideau Lakes are so formed that an exceptionally large reserve of water can be accumulated during the spring run-off and stored back for use during the low rainfall months of June, July and August. However, even with this natural reservoir a normal summer precipitation is required if the demands for water are to be satisfied. And these demands are heavy as we shall see.

The Precambrian Shield in the area immediately west of Westport may be likened to a giant saucer tilting slightly towards Upper Rideau Lake, the highest navigation point on the Rideau Canal. The large lakes within this saucer including Bob's Lake, Wolfe Lake and Westport Lake feed a tremendous volume of water into Upper Rideau Lake. From here part will flow north-west through Narrows Lock down to Ottawa and the remainder will flow in a southerly direction through Newboro Lock towards Kingston.

The rate of flow into Upper Rideau Lake, and through the two halves of the canal system is kept under careful control at many strategic locations. The first of these is at Westport where a dam regulates the quantity of water drained from Bob's Lake which is some 124 feet higher than Upper Rideau Lake.

The main control dam for the north-eastern section running from Narrows Lock to Ottawa is at Poonamalie, just south of Smiths Falls. This dam has automatic regulation controls that govern the flow rate past this point. Big Rideau Lake is used as a huge reservoir since it has no other outlet but through this control dam. Essentially, the only demand for water on this route to Ottawa is that there be sufficient to maintain a minimum five feet draught in the canal channel.

The southern section of the canal, from Newboro to Kingston, is able to control the flow rate at each of the seven lock stations. In this section are four hydro electric generating stations, one each at Jones Falls, Upper Brewers, Lower Brewers and Kingston Mills.

These four plants, especially the one at Jones Falls which has four generators, make a heavy demand upon the water supply and sometimes have to be turned off if the level of water in Upper Rideau Lake falls too low.

The electrical power generated by the four stations is fed by high voltage transmission lines to Gananoque to satisfy all the requirements of that community and the remainder is carried to Kingston, there being sufficient to provide for about half its needs. A private company operates these generating stations and co-operates closely with the Dept. of Transport in maintaining an adequate supply of water in this portion of the canal system. To do this properly it has invested considerably in means of storing water in Devils, Hart, Rock and Loughborough Lakes. The release of this water into the canal system for power purposes is of benefit to navigation as well.

Without a constant supply of water, properly controlled for maximum efficiency, the Rideau Canal system would not today be a pleasure cruiseway enjoyed by tens of thousands each year. The builders themselves recognized it was necessary to control levels and the system has been designed with overflow dams and by-pass weirs to achieve this result. Admittedly, they did not have to contend with hydro electric plants that gulp water quicker than the locks, nor the many times the locks nowadays have to be opened to permit passage of pleasure craft, but they did anticipate lean precipitation years and took measures to prevent the main channel from becoming unnavigable.

The water levels in the main canal channel and in the surrounding lakes are kept under careful control all summer long.

ℱishing

THE RIDEAU LAKES

OPEN SEASONS FOR ANGLING		LIMITS OF CATCH
Largemouth Black Bass	Jun 26 - Oct 15	6 in one day
Smallmouth Black Bass	Jun 26 - Oct 15	6 in one day
Muskellunge	Jun 26 - Nov 30	2 in one day
Brook or Speckled Trout	Feb 27 - Oct 4	10 lbs plus 1 fish, or 15 fish, whichever is the lesser.
Atlantic Salmon or Ouananiche	May 1 - Oct 15	1 in one day
Pike, including Great Northern Pike, Grass Pike and Jackfish	May 8 - Mar 31	6 in one day
Yellow Pickerel, Walleye, Dore or Pikeperch	May 8 - Mar 31	6 in one day
Lake Trout, Brown Trout, Splake	Feb 27 - Oct 4	5 in one day

Subject to Change

Minimum size limits prevail in certain districts.
Check the regulations carefully.

ONTARIO ANGLING LICENSE FEES

Non-resident (entire season)	$6.50
Non-resident (three days only)	3.25
Residents (Provincial Parks only)	3.25

Some of the best fishing to be found in Ontario lies within a fifteen mile radius of Westport, an area containing about forty-five lakes. One has a wide choice of bass, trout, pike, yellow pickerel (or walleyed pike) and muskellunge. Trout up to 48 pounds and muskellunge up to 54 pounds have been caught and, while not common sizes, serve to indicate the excellent catches that can be made.

Many of the Rideau Lakes are restocked annually with bass and trout fingerlings and pickerel eggs. This service contributes materially to maintaining an abundance of fishes in this part of the Waterway. Scientists learned long ago the waters themselves are highly conducive to the natural reproduction of certain species

of fish. The hatchery on Lake Opinicon especially has conducted
tests confirming the "population explosion" each year of the warm
water breeds. To say there is an abundance of fish in these lakes
is putting it mildly.

If you are a keen angler we suggest you enquire of the Chamber
of Commerce serving the area of your choice for guide and accommodation information. Chambers are located at Perth, Smiths Falls,
Westport and Kingston — and of course don't forget Marguerite
Stoness at Battersea. You may also write to The Rideau Lakes-
1000 Islands Tourist Council, Box 486, Kingston, Ontario for prompt
and cordial service.

Dimensions of all locks	–	134 feet x 33 feet
Vessels limited to	–	Length 110 feet, beam 30 feet
Draught on lock sills	–	Normal 5.5 feet, minimum 5.0 feet
Minimum bridge clearance	–	Rideau Canal – 22.0 feet
	–	Tay River Branch – 5.18 feet

Canal Data

Number of locks	–	Rideau Canal – 47
	–	Tay River Branch – 2
Number of dams	–	24
Elevations above sea level	–	Ottawa Lock 1 - 132 feet
	–	Kingston Mills Lock 49 - 246 feet
	–	Upper Rideau Lake (Summit) - 408 feet
Total lift	–	14 locks Kingston to Upper Rideau Lake – 162 feet
Total descent	–	33 locks Upper Rideau Lake to Ottawa River – 276 feet

Operation Schedules

The season for operation is from May to November inclusive.

All locks remain closed daily from 12.30 p.m. to 1.00 p.m. and from 5.00 p.m. to 5.30 p.m. for Canal Staff lunch periods.

The Smiths Falls Beckwith St. Swing Bridge will not operate during rush traffic periods 8.30 a.m. to 9.00 a.m., 12 noon to 1 p.m. and 4.45 p.m. to 5.45 p.m. weekdays.

The Hogs Back and Pretoria Ave. Bridges in Ottawa will not operate during rush traffic periods 8.30 a.m. to 9.00 a.m., 12.30 p.m. to 1.15 p.m. and 4.15 p.m. to 5.45 p.m. weekdays.

The Rideau Ferry, Burritts Rapids and Brass Point Swing Bridges will operate as required by boat traffic.

The annual schedule for the operation of the canal system is established each year by the canal authorities. Generally, during the summer months the locks are in operation between 8.30 a.m. and 8.30 p.m. Enquire the schedules from any lockmaster or write the Superintending Engineer for particulars.

Average normal locking time — 15 minutes per lock.

Rideau Canal Operation The maintenance and operation of The Rideau Canal is the sole responsibility of the Federal Department of Transport. Mariners and visitors to the locks are requested to obey the instructions of the Lockmasters and attendants at all times, since these are intended only for the service and safety of all concerned. The lock personnel are an exceptionally fine bunch of men and will treat you courteously at all times. Please respect their responsibilities and the fact they are subject to instant dismissal if found accepting alcholic beverages from visitors, no matter how well intended such gifts may be.

Matters concerning the operation of the canal should be referred to:

> Mr. L.W. Clark,
> Superintending Engineer,
> Rideau and Nova Scotia Canals,
> P.O. Box 902,
> Ottawa, Ontario.

Navigation Information This guide does not purport to prescribe information or instructions for navigating the channels or the lock approaches. Only the official Navigation Charts listed in this chapter should be used for this purpose.

Approach Signals Give warning of your approach with three long blasts each of 5 seconds duration.

Approach Wharves An "Approach Wharf" is to provide adequate space for vessels to tie-up while they are waiting to be locked. Each Approach Wharf is designated by a broad blue line painted along its length and a white letter "A" on a blue background. Vessels wishing to tie-up for a few hours should obtain permission from the Lockmaster of that particular station and should tie-up on a portion of the wharf other than the Approach Wharf.

Pollution of waters
Mariners are especially requested to use refuse containers provided at each lock station and to use toilet facilities on shore where available. Please do not throw garbage into the water or on the shore.

Weed Obstructions
Many areas adjacent to the Rideau Canal main navigation channel are subject to heavy water weed growth. Keep to the main channel at all times and proceed with extra caution if you must deviate from the well marked channel.

Swimming
Due to the severe undertows prevalent at each lock swimming in the vicinity of the locks is unsafe and is strictly prohibited.

Camping
Some space is provided at the majority of locks for boaters desiring to camp overnight. These are not listed in this guide as generally the spaces are limited and no promise can be made that all conditions are adequate, although the areas are kept clean and tidy and as private as possible.

Tolls
Free passage through the canal for pleasure craft is provided with the compliments of the Canadian Government.

Customs Clearance
Boats entering Canada by water from the United States are required to clear through customs. A convenient Customs Office is located at Kingston, Ontario, at the entrance to the Rideau Canal.

Publications
The following Publications and Navigation Charts are recommended for use by all persons using the Rideau Canal.

1. NAVIGATION CANALS. Published by Department of Transport. Cost $0.35 each. Order from Queen's Printer, Ottawa or Rideau Canal Office, Box 902, Ottawa, Ontario. May be purchased direct from Lockmaster at Ottawa Lock Station, Lockmaster at Kingston Mills Lock Station or from Canal Superintendents at Ottawa, or Smiths Falls.

2. Rideau Canal Navigation Charts 1575 and 1576. Cost $0.50 each.

   Order from Chart Distribution Office, Surveys and Mapping Branch, Canadian Hydrographic Service, Department of Mines and Technical Surveys, 615 Booth St., Ottawa, Ontario. May be purchased direct from the Lockmaster at Kingston Mills, Smiths Falls and Ottawa Lock Stations or the Rideau Canal Office, 340 Queen St., Ottawa. Some marinas stock these charts and the Chamber of Commerce in Smiths Falls also sell them as a public service.

3. Topography Map. Sheets 31 B, 31 C and 31 G. Cost $0.25 each. Order from Department of Mines and Technical Surveys, address above.

   Note. In Canada, remittances should be made by Money Order or Accepted Cheque, payable to the Receiver General of Canada. From the United States, remittance should be made in Canadian Funds by United States Postal Order payable to The Receiver General of Canada.

The Headquarters of Col. John By, now The Bytown Museum

MARINAS AND DOCKS

LOCATION	MILES TO NEXT FACILITY
KINGSTON	4½
KINGSTON MILLS	18½
SEELEYS BAY	5
JONES FALLS	6½
CHAFFEYS LOCKS	5
NEWBORO	10
WESTPORT	15½
PORTLAND	12
RIDEAU FERRY	8½
SMITHS FALLS	14
MERRICKVILLE	21
KARS (1 mile north)	2½
MANOTICK	6
BLACK RAPIDS	4
HOGS BACK	1½
DOWS LAKE	3½
OTTAWA	----

CANAL STATISTICS.

STATION	NO. OF LOCKS	MILES FROM KINGSTON	MILES FROM OTTAWA	AVERAGE LIFT FEET	MILES TO NEXT STATION
KINGSTON MILLS	4	4.62	118.91	44.99	9.85
LOWER BREWERS	1	14.47	109.06	13.00	1.78
UPPER BREWERS	2	16.25	107.28	18.00	10.69
JONES FALLS	4	26.94	96.59	58.50	4.44
DAVIS	1	31.38	92.15	9.0	2.15
CHAFFEYS	1	33.53	90.00	10.75	5.26
NEWBORO	1	38.79	84.74	7.75	4.72
THE NARROWS	1	43.51	80.02	3.0	19.04
(TAY BRANCH TO PERTH) 6.82 miles					
BEVERIDGE	2	57.21	66.32	25.0	-
POONAMALIE	1	62.55	60.98	5.75	2.10
SMITHS FALLS	4	64.65	58.88	34.50	1.16
OLD SLYS	2	65.81	57.72	16.0	1.5C
EDMONDS	1	67.31	56.22	9.16	3.41
KILMARNOCK	1	70.72	52.81	2.0	8.0
MERRICKVILLE	3	78.72	44.81	24.66	2.31
CLOWES	1	81.03	42.50	7.58	0.41
NICHOLSONS	2	81.44	42.09	14.50	3.16
BURRITTS RAPIDS	1	84.60	38.93	9.0	24.68
LONG ISLAND	3	109.28	14.25	25.33	5.0
BLACK RAPIDS	1	114.28	9.25	9.16	4.02
HOGS BACK	2	118.39	5.23	14.50	0.97
HARTWELL	2	119.36	4.17	21.50	4.17
OTTAWA	8	123.53	0.00	79.0	-

Profile of Canal

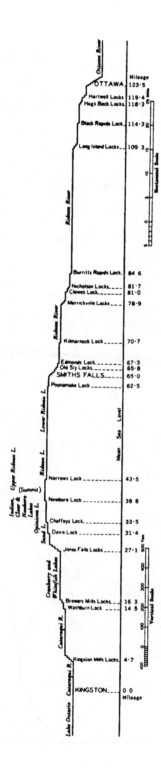

	Mileage
Ottawa River	
OTTAWA	123·5
Hartwell Locks	119·4
Hogs Back Locks	118·3
Black Rapids Lock	114·3
Long Island Locks	109·3
Rideau River	
Burritts Rapids Lock	84·6
Nicholson Locks	81·7
Clowes Lock	81·0
Merrickville Locks	78·9
Rideau River	
Kilmarnock Lock	70·7
Edmonds Lock	67·3
Old Sly Locks	65·8
SMITHS FALLS	65·0
Poonamahe Lock	62·5
Lower Rideau L.	
Rideau L.	
Upper Rideau L.	
Indian, Clear & Newboro Lakes (Summit)	
Opinicon L.	
Narrows Lock	43·5
Newboro Lock	38·8
Sand L.	
Chaffeys Lock	33·5
Davis Lock	31·4
Jones Falls Locks	27·1
Cranberry and Whitefish Lakes	
Brewers Mills Locks	16·3
Washburn Lock	14·5
Catarqui R.	
Kingston Mills Locks	4·7
Cataraqui R.	
KINGSTON	0·0
Lake Ontario	Mileage

Mean Sea Level

Horizontal Scale

Vertical Scale

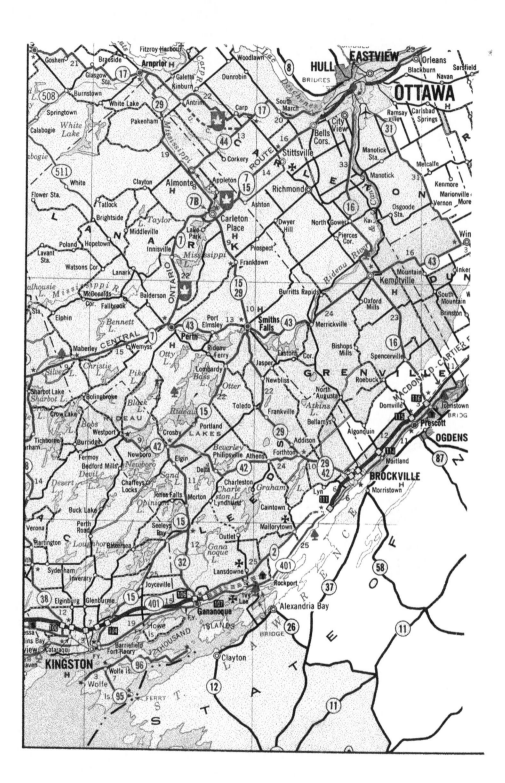

F
5508
R5

Rideau waterway guide

WS - #0171 - 051222 - C0 - 229/152/4 - PB - 9781332859948 - Gloss Lamination